Leon Mead

Sky Rockets

Leon Mead

Sky Rockets

ISBN/EAN: 9783337037857

Printed in Europe, USA, Canada, Australia, Japan

Cover: Foto ©berggeist007 / pixelio.de

More available books at **www.hansebooks.com**

SKY ROCKETS

By LEON MEAD

BOSTON
PUBLISHED BY THE AUTHOR
1883

CONTENTS.

DEDICATION.

	PAGE.
To my Mother,	5

MISCELLANEOUS.

The Poet's Satisfaction,	7
An Autumn Idyl.	9
The little Red School-house,	11
The Sequel of Life,	13
At Rest,	14
A May Morning,	16
Shall I see them no more?	17
The Good that is,	19
Laune,	21
Unsullied Faith,	24
In a Grove,	26
Mohammed's Heaven,	28
Longfellow,	34
A Souvenir,	36
The Farmer's Life,	37
October,	40
The Coming Era,	42
Work,	46
Dusk and Dawn,	48
Affinity, a Romance,	50

CONTENTS.

SONGS AND VAGARIES.

The Cascade of Romance,	68
Remorse,	70
Lady Rose,	72
A Bachelor's Story,	74
The Same Answer,	76
Her Name,	78
Love's Silence,	79
A Maiden Lady,	80
Redeemed,	82
Write often to the Old Folks,	84
An Ideal Home,	86
A Peace Offering,	88
Maraschino,	90

LEVITIES.

Sonnet to an Organ-grinder,	93
Triolets,	94
Only Cousin Sam,	96
The Reason,	97
In 1901,	98

TO MY MOTHER.

Her gentle presence lingers now around me,
   The quiet balm of tenderness she shed,
Is still a power in memory to cheer me,
   When all the world to me seems cold and dead.

For other friends whose confidence I cherished,
   Affection's spray has frozen in my heart;
Since its refreshing fountain sadly perished,
   When they refused to longer play their part.

She could excuse the irksome duty,
   She could forgive recurring wrong;
There was always meekness in her beauty,
   There was always beauty in her song.

Away from home her spirit travels with me,
   As though my erring ways it understood,
And would from evil dangers always lead me,
   And pilot me where ev'rything is good.

A guide to higher aims than simply pleasure—
   A counsellor for God, a faithful lover;
The one to tell my secrets without measure,
   The one to ask for favors was my mother.

God bless the women that are like my mother,
   Who makes our home by love a Paradise,—
Ah! in my eyes she e'er has seemed no other
   Than but a perfect angel in disguise.

MISCELLANEOUS POEMS.

THE POET'S SATISFACTION.

They only know the mental cost
   Of weaving all the woof of song,
Who still, when trenchant youth is lost,
   Retain the loom they've worked so long.

The tinsel garbs of poesy,
   With fairest, sweetest flowers enwrought,
But only hint the truth to me
   Of what they felt and thought.

The poet with the snow-flecked hair
   Of sunset scenes grows strangely fond —
He sees new beauties glowing there,
   Suggestive of the life beyond.

While he e'er radiates his fire
    To each far clime and distant zone,
The world converges with desire
    To pay its tribute at his throne.

Proud are the trophies of his toil
    Weened from the models of his youth;
The art flower bloomed in nature's soil,
    And withers not its scent of truth.

Verses that he penned for naught
    But pleasant pastimes, as he said —
Have more than idle fancies taught,
    And they will live when he is dead.

AN AUTUMN IDYL.

The presence of autumn is here;
    The daisies are drooping and dying,
The blood of the arbute is chill,
    The leaves of the maple are sere,
And thro' them the breezes are sighing,—
    The presence of autumn is here.

The sadness of autumn is here;
    The brightness of summer is over,
The grass on the hillside is wan,
    The sky is portentous and drear,
The bees have forsaken the clover,—
    The sadness of autumn is here.

The muteness of autumn is here;
    The swallows have hastened away,
The winter birds sing on the bough
    A song that lacks music and cheer;
All nature seems wrapt in decay,—
    The muteness of autumn is here.

AN AUTUMN IDYL.

The beauties of autumn are here;
    The verdure is gilded with gold,
The foliage blushes with crimson,
    The clouds dipped in sunset are clear,
The orchard is rich to behold,—
    The beauties of autumn are here.

The lessons of autumn are here;
    As told in the Storm King's harsh breath,
Which warns men how fleeting is life;
    That at the lone grave of the year
The flowers are lovely in death,—
    The lessons of autumn are here.

THE LITTLE RED SCHOOL-HOUSE.

In the grave of the past it is buried,
   That weather-worn temple of wood;
And only dank weeds in the summer
   Now mark the dear spot where it stood.
Oh! could all the scholars assemble
   Once more in that prison-like place,
And hear the quaint schoolmaster utter
   His heartfelt entreaties for grace!

Oh! could we return to that school-room,
   Untouched by the evils of years,
And find the bright smiles that have vanished
   In place of the dimness of tears,
And join in the silver-toned laughter,
   The gurgle of innocent fun;
The races we had going homeward
   When all the hard lessons were done.

How the truants sulked in after bell-time,
   So guilefully heedless of rule;
For they knew the old teacher was patient,
   His smile was the law of his school.

What a scene was this beehive of knowledge
   On hot, murky days in July,
When the little ones turned from their studies,
   In the shade of the elm tree to lie.

But alas! the crude structure has fallen,
   Its timbers have gone to decay;
The master sleeps there in the corner,
   Where the glad children shouted in play.
God bless the dear spot that since childhood
   Has grown to be sacred and still,
Where the little red school-house in glory
   Once stood on the brow of the hill.

May all of the scholars assemble
   In Heaven's great class-room above,
And meet after life's fitful season,
   To learn the grand wisdom of love;
And see the old docile-faced teacher,
   A pupil himself as before,
In branches whose worth he commended
   In the little red school-house of yore.

THE SEQUEL OF LIFE.

The bright ambers glow in the ocean below,
   And wisdom's the pearl on the mind's pebbly shore;
Could we happiness gain quite as easy as pain,
   Each heart would be filled with some valuable ore.

Strange alchemies weave the fair sunset at eve,
   Ere the gems on the grass are concealed by the night;
Ah, such beauty untold is not purchased with gold,—
   But the poorest of men may be blessed by the sight.

The rainbow above is God's symbol of love,
   And the spirit of nature invites our communion;
Hearts blend into one like the motes in the sun,—
   Death itself is unable to sever their union.

Some driftwood appears on the tide of the years,
   To claim only homage which hastens his fate;
The first road to fame is an innocent name,
   And only the good can forever be great.

AT REST.

From lofty spires on every hand
  Was wafted slow the dirge of death;
The news was breathed to all the land,
  Where watched an ear for lightning's breath,—
    That he for whom our prayers were said
    At length was numbered with the dead.

Oh! why did such a requiem toll?
  'Tis not for men to question why;
We mourn the tender, noble soul,
  But God knew best his time to die.
    How peaceful that September night
    When Garfield's spirit took its flight.

His life receded like the tide
  That tumbled near the cottage door,
While silent mourners at his side
  Could scarcely think he was no more.
    Oh! touching scene for years to be,—
    Our chieftain's calm death by the sea.

AT REST.

Dead martyr of a peaceful reign!
 "A priceless treasure of regret,"
Thy memory shall be the fane
 Of liberty's proud worship yet;
  And ever at thy cherished tomb
  The immortelle of love will bloom.

A MAY MORNING.

Awake! awake! O sleeping sons of earth,
Behold the morning at her glorious birth!
The pending dewdrops string the grassy blade,
And glisten on the hill and in the glade.

The swollen brooklet, leaping on its way,
Kisses the hanging bush in eddying play,
And throws from point to point a winding sheen,—
An azure path amid the meadows green.

Upon its dewy bank, the shining flowers
Blossom the meadows into fragrant bowers;
And on the lawns beneath the spreading trees,
The lengthened shadows flutter in the breeze.

Through the pale glamour of the morning mist,
Shines the great orb from skies of amethyst;
Breathes there a balm thro' lips of vernal May,
Upon a breeze which bears it all away.

Oh truth! on such a morn must thou e'er spring
Anew in ev'ry human heart and bring
New lessons and new thoughts which speed
From Heaven high to wake the future deed.

SHALL I SEE THEM NO MORE?

A PARAPHRASE.

Shall I see them no more, must I die far away
From all I so loved in life's earlier day?
   The parent who taught me the lessons of truth,
   The brother who shared all the joys of my youth;
The one gentle sister whose smile could destroy
All the fanciful griefs of a passionate boy;
   My schoolmates so happy when study was o'er,—
   Shall I see them no more, shall I see them no
      more?

Shall I see them no more, the mountains that rose
Thro' the warm summer sky to the region of snows;
   The valley where often I pensively strayed,
   The brook where I fished and the woods where I
      played,
The house that stood near the murmuring rill
Which ran with emotion from springs on the hill;
   The maple and cherry trees near by the door,—
   Shall I see them no more, shall I see them no
      more?

SHALL I SEE THEM NO MORE?

O! bright are the skies that hang over me now,
And cool is the breeze to my feverish brow;
   I fly to the lovely and mirth-moving throng,
   I join in the laughter, the dance and the song.
But gazing on visions of beauty and grace,
The shadow of sadness steals over my face;
   I sigh for the lost ones time cannot restore,—
   Shall I see them no more, shall I see them no more?

O, God! let me die where I first drew my breath,
With my friends and my kindred around me in death;
   Let not the rude hand of the stranger be laid
   On the cold silent image of clay thou hast made.
When my spirit is gone let my body repose
In my old mountain home where the evergreen grows;
   Then those who still love me my loss will deplore,—
   Shall I see them no more, shall I see them no more?

THE GOOD THAT IS.

This world is good enough; only some of the people in it are bad.—*Old Proverb*.

In life's darkest night
Gleams a soul-cheering light
In the far away regions of bliss,
And they are not worth
A blossom-strewn earth
Who find not the treasures of this.

The sinner may prate
Of his Christianless fate;
The Christian may do things amiss;
But they are not worth
A Heavenly earth
Who see not the goodness of this.

The cynic may sneer
At the heart-sigh and tear,
And scorn a beloved one's kiss;

But he is not worth
A love-blooming earth
Who feels not affection in this.

The critics may rant,
Over art may descant,
Or at God's own creations may hiss;
But they are not worth
A beautiful earth
Who know not the beauty of this.

LAUNE.

There never was a meadow where love refused to bloom
  Among the clover blossoms which were wedded to the dew;
The frigid wind may blight the flowers and build their autumn tomb,
  But still from nature's nursery they e'er will spring anew.

There never was an ocean whose wild, tumultuous roar,
  Ne'er blended with the music of the mermaids' lonely cry;—
Love surges with the billows high upon the wreck-strewn shore,
  And rides upon the thunder as it rolls along the sky.

There never was a cradle where an angel baby lay,
   But some heart was kindly throbbing for the treasure sleeping there;
The household may be darkened by some sorrow ev'ry day,
   But the light of infant purity relieves maternal care.

There never was a churchyard without its crumbling dead,
   Who once had only lived to love, yet e'en had loved to die;
The evils of the world may cast weird shadows where we tread,
   But still men reunite in death where side by side they lie.

There never was a desert without some vernal spots,
   No human life so barren but can boast one tender dream;

Time cannot rob love's garden of its sweet forget-
 me-nots—
Though the fairest lilies wither on the bosom of
 the stream.

The magic spell which guides divinely ev'ry form
 and race,
No fancy can approach in flight, no poet's pen
 explain;
We feel and see its beauty, we adore its winsome
 grace,
And find its bliss much deeper than its very
 deepest pain.

UNSULLIED FAITH.

Whate'er may be another's moral need
To just complete his panoply of right,
   Still may I lack what he at first possessed.
A man's own life is better than his creed!
If it be pure in his Creator's sight,
   No formal faith can bring him sweeter rest.

The wretch arraigned by justice at the bar,
Is tried upon a charge of awful crime,
   And there his merits and defects are weighed;
The door of character is pushed ajar—
The verdict "guilty" rests upon his prime,
   Although he's found a churchman old and staid.

The duty lies with you and not your priest!
Your pew in church will never save your soul;
   It is by dint of worth you worthy prove,
And your good deeds will not afford the least
Of other reasons why you'll reach the goal
   Where all in all is perfect bliss and love.

But if we can our holy feelings share
In sympathy sincere beneath one roof,
   Untainted by the subtle charms of art,
Then lighter seems the mutual cross we bear—
If not, 'twere best to keep ourselves aloof
   From crafty fashions that enslave the heart.

IN A GROVE.

Here in this mossy nook, in tangled mood,
   I sit me down, while yet the sunlight streams
Its richest flood of glory thro' the wood,
   And bathes the rippling grasses at my feet,—
   To see if aught can yield me pleasant dreams,
Or lead me to reflect on something sweet.

An oak-tree shields me from the summer glare,
   And in this fresh, cool shade, the silence breathes
A dulcet song soft as an angel's prayer;
   The roses lose their pomp, and ope their lips
   To add one strain, while murmur low the leaves,
And every breeze into the music dips.

So pale and tawny look the lilies there,—
   Like thoughtless ghosts of some long-murdered flower;
Methinks they'd steal the sumac's blood, nor care

IN A GROVE.

If they possessed the green fern's stiff back
    bone,
Ah, lily frail! thine is a soulful power
Which lies in chastity, and that alone.

The broken path with poplar-trees is lined,—
    The path which to the cradled valley leads:
Upon the sweetly autumn-scented wind
    The maples toss their load of scarlet plumes;
    The wild bee on the dreaming poppy feeds;
    Then off he flies to plunder other blooms.

A plashing brook hides 'neath an alder screen,
    Where busy swallows built their summer nest;
Undimmed by tears, on such a sylvan scene,
    Heaven's blue eye must gaze with joyful glow;—
    Since it expresses love and life and rest
    For which we all are striving here below.

MOHAMMED'S HEAVEN.

I.

This tale the Prophet's scribes were wont to tell
The wand'ring pilgrims that to Mecca came,
Or those who lingered in the palm tree's shade
About Medina, near to Aysce's grave,
To which Mohammed, after death was brought—
Of his ascent up thro' Celestial spheres,
Where summer kept the promise of the spring.
They told how in the gloaming that was come,
When deep'ning dusk to slumber called the birds,
While he was kissing Aysce, his young wife,
The angel Gabriel suddenly appeared,
Clothed in a rugged flame of seething fire—
And thus saluted him: "All hail, loved man!
I come to tell thee that thy all-wise God
Whom thou adorest—He would have thee dwell
In those fair mansions of the saints above."
"But how shall I ascend?" the Prophet asked,
Still toying with the locks of Aysce's hair.
The angel bade him mount meek Albarack,
Which at four steps the holy city reached;

MOHAMMED'S HEAVEN.

And from Jerusalem was borne aloft
The Prophet to the skies, where Gabriel,
Who held, for sixty thousand gates, the keys
Of Paradise, let in the worldly man.

II.

He passes seven Heavens in a trice—
Though centuries it takes to traverse them;
The first is of bright silver most refined,
In which the stars are fixed with chains of brass
Smooth burnished as the bosom of a lake.
The second is of gold, tried many times
In fire; the third of pearl, and here he sees
The fallen Adam, Enoch and the rest.
They welcome him while he kneels down in prayer.
The fourth is of smaragdis, where appear
Still larger bands of Seraphs chanting praise;
Among them, Mercy's angel, Phatyr stands,
Holding the pen with which God registers
All mortal things, past, present and to come,
In such deep mystic characters that none
Save He and Seraphiel understand.
Now in a realm paved with glittering gems,

He meets the wisest of all angels there ;
The one who keeps the book wherein are found
The names of all men that were ever born.
The sixth is turquoise ; seventh alahab
Or fire and light by alchemies congealed.
All these enclose each other like a pearl
Or onion layers, in transparent gold.
The Heaven of all Heavens lies above,
Where light and peace and silence dwell serene.
Mohammed sees the highest throne upheld
By seven angels, on it seated God,
Who lays his hand upon the Prophet's face
In token of his love and welcome kind ;
Whereat Mohammed blushes deep with shame,
And even quakes with almost childish fear.
With full instructions he returns to earth,
Bringing the Koran filled with mysteries.

III.

The angels, he asserts, are good or bad,
Both subject to the King of Terrors—Death.
Bad angels are imprisoned in mad dogs,
And swine and toads and reptiles poison-fanged.

After doomsday they will tormented be
A million years in hell and at the end
Shall once more be delivered from their pain.
The general judgment is as certain
As the day of death, and will haply come
When this proud world may least expect it, too.
The angels do not know the day till he
(Mohammed) gives the fatal sign to men—
A fearful duel 'tween Adriel and Death,
Who, in the end, is vanquished yet so 'raged,
That he destroys all living things below,
And then ensues an earthquake, violent,
A shower of purling brimstone following,
Which kills the grass, the trees and lovely flowers,
And turns earth back into its first wild Chaos.
For forty days in this sad state 'twill be,
When God will rain for forty days and nights
A shower of tender mercy down and then,
By one sweet, gracious, gentle breath reduce
The world into a glorious estate.
He will call Seraphiel, bid him blow
A trumpet, at whose sound angel and man
Shall wake, the one his glory to resume,

The other his flesh. In his rude balance
Michael shall poise the deeds and conduct
Of men, and they whose good traits overpoise
Their bad, are placed in joy at the right hand—
The others in their guilt upon the left.
Those laden with their sins are to pass o'er
A narrow and weak bridge to endless shades;
Such as possess few sins in safety cross,
While those of many break the fragile piers,
And fall into perdition's dark abyss.

IV.

But of the nature of this Paradise;
It is a place of such delight, indeed,
As but Mohammed's mind could picture it.
He thinks 'twill be upon the earth when all
The dead are raised and judged and doomed.
It is divided into seven groves
Or spacious gardens where supernal joy
In countless places of enchantment reigns.
Each one is filled with laughing damosels,
Who dance for wine to music soft and sweet;

Each bower's refreshed by cooling rivulets
Of crystal dribbling over precious stones;
Aromatic flowers abound which ravish
Eye and smell; the fond birds please with far more
Volup'ous song than sweetest nightingale;
The airs compounded of nectarous scents—
Mohammedans, the Christians or the Jews
Know not such happiness, are ever young
And spritely there; the amorous women
Lie in charming groups beneath the lilac boughs,—
More sensuous grows the scene till we would fain
Retreat as from the sight of some vile beast,
And seek for refuge in the path of faith,
Which our fond mothers taught us in our youth,
Would lead to lofty realms where the soul
Can share the exaltations of its god.

LONGFELLOW.

Our Poet-King has left his throne,
   And all the muses touched with pain,
Ask. mind their tears in solemn tone,
   "When shall we see his like again?"

Hushed is the velvet voice, whose spell
   O'er friends a charm could e'er impart;
Still is the pen which, like a bell,
   Could chime the cadence of his heart.

And rigid is that furrowed brow,
   Which mirrored all his genius bright;
And silent are the pulses now
   Which beat in harmony with right.

But, hushed is not his country's sigh,
   For that great presence it has lost;
Oh! they are always sure to die
   Whom we appear to need the most.

His breathing words death cannot still,
　　They grace a grand immortal scroll;
His burning thoughts death cannot chill—
　　For they are sparks from his own soul.

Ah! rigid not is that blest soul
　　Which lately joined the angel throng.
Whose anthems now will deeper roll
　　In welcome to the Prince of song.

Nor silent will his spirit be,
　　Released from flesh beneath the sod;
For thro'out all eternity
　　'Twill be a favored guest of God.

A SOUVENIR.

On a once priceless morn, thro' the valley I strolled,
  To say a farewell to the fond scenes of yore;
At my left the dark waves of the Delaware rolled,
  On whose bloom-covered banks I could linger no more.

And I thought of my friends and the future ahead;
  What changes would come e'er that summer should close,
If fate would announce that some dear one was dead,
  And I plucked from the bush a bud of wild rose.

I cling to it still, though 'tis withered and frail,
  As a keepsake of all that was dear at life's dawn,
To recall the last time that I strayed thro' the vale,
  Enwrapt in a dream of the summer that's gone.

THE FARMER'S LIFE.

This poem was delivered at a Farmers' Annual Picnic by the author.

The muse has been a friend to men of toil,
   Since first from high Parnassus' lofty grove,
   Her harp was tuned by Heaven's choir above,
To cheer them on in labors of the soil.
   And here we meet to-day, a rural band,
To celebrate the progress of man's noblest aim—
   In nature's sylvan shade and Freedom's land,
Where farmers oft have kissed the lips of fame.
   The pride of place awaits the farmer now
   As when a Roman for it left his plow.

To this calling then should we aspire,
   Who know not this world's honest gain;
   But strive for more than we attain,
And die poor victims of some vain desire.
   The kindly voice of nature bids us sow

The virtuous seed of labor in the field,
    And check malicious weeds that only grow
To choke the blossoms that would gladly yield.
    The pride of place awaits the farmer now
    As when a farmer for it left his plow.

To you who feel a poignant sense of pain,
    That life upon the farm is full of grief;
    The sight of urbane sorrow brings relief,
And leads you to denounce the pride as vain.
    The hand for work and duty's willing slave,
God gave to man perfection to essay;
    E'en imitate our fathers in the silent grave—
They left us the foundation of to-day.
    The pride of place awaits the farmer now
    As when a Roman for it left his plow.

The blushing maid all innocent of shame,
    Whose birthplace was among these rustic bowers,
    Is worthy of a prince's royal dowers,
Far more than one who worships fashion's name.
    She may be called a mockery of art,
By those whose pencilled brows at plainness frown;

But when it comes to sweetness of the heart,
She well could be the belle of any town.
A glory waits each maiden in the land,
As when Joan of Arc assumed command.

OCTOBER.

I.

We often sigh when the pansies die,
    Though autumn yields her dower;
There is dreaded woe in the pallid glow,
    And the death of a single flower;
Yet why should we grieve while nature lies
In the lap of October's paradise?

Through an open rift in the fleecy drift
    Fade the radiant lamps of night,
O'er the garden wall the sunbeams fall
    In a tangled mesh of light;
And they drink from the sullen stream below,
Where a clump of wild thyme used to grow.

II.

Strew immortelles along the way
    Where wan a million petals seem,
The harvest field is gleaned to-day,—
    And willows whisper to the stream,
    That all the past is but a dream.

OCTOBER.

The latest swallows southward fly,
    Because the leaves to russet turn;
No more they cleave the summer sky,
    Nor warble 'mid the meadow fern,
    And for them now 'twere vain to yearn.

A lesson has the year's decline,
    As crimson poppies slowly swoon;
The rose is neither yours nor mine—
    It serves a purpose like the moon,
    Nor fades until it does some boon.

What secrets linger round the tomb
    Where nature sleeps in death's deep calm,
Till spring will resurrect her bloom,
    And heal the winter wounds with balm
    Plucked from the earth's own vernal palm.

Let chill October's lyre repeat
    The dirges sad which vibrate loud;
A few stray sunbeams yet may greet
    The snowflakes fleeing from the cloud,
    And fringe with glory winter's shroud.

THE COMING ERA.

Low in the dust and silence, low in earth's virgin
    breast,
  Clammy and cold and senseless there in their
    slumber deep,
The victims of drink are lying in a mute and soul-
    less rest,
  And sealed are the lips that dying asked for a
    peaceful sleep.

Grasses wave above them and heavy with twilight's
    tears,
  The roses wan and weary lean over the vernal
    slopes
To hear the spirit voices that come from the by-
    gone years—
  That speak of human ruins and the ghost of
    murdered hopes.

They tell of the mystic shadows that crouch by
    hearths aglow,
  Where wives are sobbing wildly and mothers
    sigh in pain;
Where dregs of bitter memory fill up their cup of
    woe —
  Where all their prayers for lost ones are uttered
    but in vain.

Of manhood's deadly grapple and subsequent defeat,
  With one weird dancing demon in sulphurous
    habit decked;
Of merry hearts that drifted out on billows wild
    and fleet —
  Of hearts dashed on sin's hidden reefs, of hearts
    forever wrecked.

And the life and love of many a home have gone
    to the distant skies,
  Like mist that coils from the river, or the incense
    of battle's fray;

Oh! hard is the lesson we gather when the drunken
    parent dies,
  Leaving the curse behind him, perhaps to his
    children's day.

The wail of the orphan is drowned in the ceaseless
    din of the street,
  While rivers of wine flow down the throats of
    the rich and the proud;
And rampant the evils and crime which everywhere
    we meet,
  And the shuttle of death keeps weaving the poor
    inebriate's shroud.

Reeking the cells of the prisons with the poison
    breath of the wretch,
  Filled is the almshouse with paupers and tramps
    tattooed with shame;
Souls are pawned for a trifle, and honor for what
    it will fetch,
  And duty bleeds with wounds she receives in
    pleasure's name.

THE COMING ERA.

Over the Empire, Progress, calm as the stars above,
  Rides in her chariot golden, urging her charger's time;
The banners of Heaven floating the gilded message of love—
  Inscribed thereon by the angels to men of every clime.

And ne'er will she pause in her journey along the future's track,
  Till dram shops are changed to mansions where joy and love can dwell;
When souls are redeemed, homes restored, and the virtues of men come back—
  Ah! then will she restfully pause and say to all our land, "It is well."

WORK.

In the workshop, in the foundry,
   In the caverns of the ground,
Labors there a mighty Cyclops,
   Through the sun's eternal round.

And the busy marts of commerce,
   And the iron roads of trade,
Ever echo with the tumult
   By his ceaseless labors made.

From the throats of crowded fact'ries
   Rises dark the dust and smoke;
And the hammer-beaten anvil
   Knows no respite from the stroke.

Time has passed in years unnumbered,
   Cities he has built and raised;
At mutation's vastest temple,
   Nations have in wonder gazed.

WORK.

Hast thou looked upon his labors,
   Hast thou seen how he has wrought?
At the forge and at the anvil,
   Days with truth eternal fraught.

Fear not brother at his toilings,
   Truth triumphant guides them all!
Deep he labors and in secret,
   But his plans shall never fall.

Trust thou then this strength, eternal,
   Toiling in the smoke, to sign
Labor charters of the ages
   Waxing glorious thro' time.

DUSK AND DAWN.

Crystal snow the landscape covers, over all the
    twilight hovers,
  Like a mourner o'er a bier;
There is cause for nature's sadness, there is none
    for human gladness,
And 'tis well to just remember, this the last day of
    December
  Is the last day of the year.

He is old and thin and hoary, all his griefs and all
    his glory
  Will be buried ere the dawn.
While the cross and crown he carried, many died,
    and some were married —
Only twelve months back he mounted in his youth
    Time's throne ghost-haunted
  By the old years dead and gone.

He was often gay and cheerful and he robbed the
    sad and tearful
  Of their sufferings ev'ry day;
Though he sometimes, without reason, seemed to
    foster sin and treason —
Yet, the dirges which March chanted, he so willingly
    supplanted
  By the robin's song in May.

We shall wake upon the morrow, to our happiness
    or sorrow
  With a new King on the throne;
Ah! his heart has had the schooling, that's pre-
    pared his mind for ruling,
And the New Year's heart snow-chastened, may by
    ties of love be fastened
  Close forever to our own.

AFFINITY.

PART FIRST.

The mansion stood upon a terraced green,
    And, down below, the Susquehanna sped
In languid motion like a stately queen,
    Or like a cortege moving with the dead.
The porch was shaded by a trellis screen,
    And morning sunshine filtered thro' and shed
A fragrant glory in the *boudoir*, where,
Amanda sat, in all her beauty fair.

And here she mused, for in her life was naught,
    In all its joyance, half so sweet to her,
As what she by her subtle fancies wrought
    Of ideal epochs that would ne'er occur;
But yet as secret yearnings they were fraught
    With transports which her inmost soul did stir—
Her gold was dross to all her wealth of thought;
Her vagaries were neither sold nor bought.

Sometimes she trembled lest she should betray
  By look, or speech or action what had been
Her fondest dream thro' each prosaic day—
  As though the habit were a ghastly sin
To mould Apollos out of common clay—
  Those to the awful canaille not akin,
And court them in coquettish modern style,
Instead of aping old romantic guile.

We like to have about us what we lack
  To make ourselves in human gifts complete;
If we could into just one person pack
  The virtues of the race, since Adam's feet,
Thro' Eden's shade trod in the serpent's track,
  And all men's genius, strength, in him should meet—
Methinks it would be hard to draw the line
Where he was much less human than divine.

Some beauteous phantoms lurk in ev'ry one,
  In dim recesses of the soul, unknown,
Until called out by love's enticing sun,
  They caper round our mind's fantastic throne;

Cheering our moods more than our friends have done,
    Growing so tangible that we are prone
To half expect affinities to find
In all the sylphs that populate our mind.

No one of all the cavaliers, who came
    To tempt Amanda with their offers grand,
Could kindle in her love's consuming flame;
    Much less could these poor fellows understand,
Why she displayed so much of scorn and shame
    When they entreated for her heart and hand.
She was a human solitaire that stood
The brightest in the crown of womanhood.

Her tender years were passed in orphanage,
    And then at Vassar she was duly sent—
To study Greek, and make herself a sage,
    And after graduating, 'Manda went
Abroad, where art and music are the rage.
    In royal courts she was an ornament;
But tiring of this foreign glamour, she,
Returned, an heiress, to large property.

AFFINITY.

She scarcely knew why she was mingling still
   In social life with all its sycophants;
Or why she sought one precious hour to kill,
   In flippant gossip or in whirling dance,
When dreamy solitude could ever fill
   Her spacious soul with a delicious trance
Beyond the world's dull, morbid cant and caste,
Which darken all the pages of the past.

First came a dainty ripple, then a flood
   Of tinted passion, into Russell's face,
When he beheld Amanda, as she stood,
   The very image of angelic grace;
Charming her neighbors with her plastic mood.
   The *soiree* would have been so commonplace,
Had she not lent to it a sort of bliss,—
At least, her new admirer reckoned this.

He vowed a modest overture to make,
   And gain at once Amanda's high esteem;
His humor was vivacious—sure to take,
   Albeit, George too sober oft did seem.

He studied themes that others would forsake ;
  His smile was passive as a frozen stream—
And sometimes in his utter dreariness,
He longed the fine emotions to possess.

To him, the sun, the stars, the sapphire sky,
  In all their splendor made a mute appeal ;
But for his warm embrace they were too high,
  And he must be content to only kneel.
Then by some mundane contrast he would sigh,
  That like a Pagan he so much did feel ;
Men go in quest of prizes far and wide
When just the thing they want is by their side.

He had a boding of the darksome eve
  When shiv'ring with a philosophic chill,
In bleak retirement, he must surely grieve,
  Because the world had treated him so ill ;
Because it would not give him kindly leave
  To patronize the highest circles still,—
We're in the very worst of prisons, when,
Within, ourselves, we can't get out again.

He felt the lack of something in his life—
    A something that would fire his very soul,
And urge him on into the worldly strife,
    Until he reached ambition's cherished goal.
He thought a sweet, devoted, little wife
    Could play for him the leading lover's role;
And be of comfort at some future day,
When all his raven locks were changed to gray.

With falt'ring step, George to the mansion came,
    One lovely day—a perfect day in Spring;
And tremblingly he broached Amanda's name,
    To John, the servant, answering his ring.
Then he was duly ushered by the same,
    Into the drawing-room, where ev'ry thing,
Indeed, was so magnificent and grand—
He fancied he had entered fairy land.

The statuary, standing here and there,
    The virgin faces, types of innocence;
The bronzes with heroic bosoms bare,
    In attitudes of virulent suspense;

The paintings, ceramics and vertu rare—
  The mirrors flashing out the hues intense
Of glowing Persian rugs and carpets gay,
From magic looms in India far away.

The chandelier of vari-colored glass,
  The centre table loaded down with books;
The frescoed ceiling, filagree of brass
  That decked the alcoves and the tiny nooks
Filled with seaweed, and shells and sallow grass,
  Above the grate—upon all these he looks
In silent wonderment and smiling awe—
As though the like before he never saw.

At last, he turns and, near a *lunette*, sees
  The image of Amanda sitting there,
In Southern, sumptuous, delightful ease,
  Upon a richly cushioned easy chair;
Ensconced in most bewitching draperies—
  A simple rosebud fastened in her hair.
She closed her novel with a pretty frown,
Then bowed, and bade her visitor sit down.

George half reluctantly picked out a seat
   Three feet away, where he could faintly smell
The odor of that rosebud, oh, so sweet!
   Blent with a sweeter fragrance, he knew well,
Belonged to those brown tresses ever neat.
   Just why, 'tis difficult for me to tell,—
But that soft perfume of a woman's hair,
Has made good men the martyrs of despair.

Their conversation was quite dull, at first,
   Until Amanda deftly broke the ice,
By speaking of her novel as the worst
   That she had read since coming back from Nice,
Where ev'ry one so frequently rehearsed
   Fine passages from Ouida, that she twice
Essayed to read, since people said she must—
But had left "Moths" unfinished in disgust.

Thus offered a fair chance to show his taste
   For mental products of the better kind;
George passed from age to age in rambling haste,
   Revealing something of his depth of mind;

Now going cross lots o'er time's awful waste,
　　Then back to some poor genius left behind,—
Concluding that he liked Shakespeare the best,
Amanda listened with great interest.

For most young men in these disastrous days,
　　Can gossip well on horses and prize fights;
And bicycles, and blizzards, and new plays,
　　And yachts and roller skating, woman's rights,
And famous refugees from England's haze,
　　And doings at the club on gala nights—
But on the wisdom of old Socrates,
They know no place wherein a word to squeeze.

George quoted Sappho's ode to Aphrodite,
　　And other remnants from her lyric pen,
At which Amanda showed a calm delight;
　　Because he seemed so different from men.
And when he paused, reflectively, a mite,
　　She asked him to repeat the ode again.
It was upon the turning edge of eve,
When George arose and, bowing, took his leave.

AFFINITY.

Oh, yes! he came again, times not a few,
    He "dropped in" at odd moments, when to call
Would have been most unseasonable, too,
    For any one not intimate at all,
With her 'bout whom young men made much ado,
    And tried in vain to scale the formal wall
That kept the world outside to emulate
The one who entered thro' an open gate.

Now, begging your indulgence, reader dear,
    I'll say that they were married in due time,
Without describing, what I really fear,
    Would get me tangled up in senseless rhyme.
Unrav'ling details in love's strange career,
    Is, I confess, a labor too sublime.
Their courtship was synthetic by the way—
A style that is quite obsolete to-day.

Of course the wedding was a great event—
    'Twas well discussed for weeks ere it occurred.
And afterwards, for weeks, Dame Rumor sent
    Fresh messages, which idle comment stirred;

Nor were the many curious content,
  Until the small particulars were heard.
Away they went the honey-moon to pass—
They thought they loved each other, but alas !

PART SECOND.

The world is wide and full of places fair,
  But give me summer in a sylvan vale,
Where winds the silent, flowing Delaware
  Upon whose moonlit bosom you must sail,
If happy fortune ever leads you there.
  The village, on its banks, is Mapledale,
And all around the Catskill mountains rise,
Great towering masses, halfway to the skies.

Below the village near the public road,
  There stood a modest farm-house in the shade
Of chestnut trees, that bore a rustling load
  Of jagged leaves, thro' which the breezes played;
The lowland meadows had been lately mowed,
  The hay into large beehive stacks was laid;
Across the field, a maiden strayed, alone,
Pausing, anon, to rest on some pet stone.

AFFINITY.

For Olive knew each appletree that stood
    Down in the orchard, where the cattle grazed;
Each meadow rock, and moss-bed in the wood,
    And sodden brooklet by the sunlight glazed;
Her aunt had seen eight years of widowhood,
    And having, of her own, no children raised,
Cared for her niece in a maternal sense,
While Olive was all love, obedience.

You see, Amanda, whom we left a bride,
    Was Olive's sis—in short the two were twins;
To mention this before I might have tried,
    But meters of this kind are full of sins,
At least, I mean, mine are, and then, beside,
    There's no use crying o'er the might have beens;
If I should start this tale anew, I trow,
It might be even worse than it is now.

These sisters looked alike and seemed to be
    Two equal halves of one all perfect thing;
In size and form, indeed, you could not see
    The slightest variance, by measuring.

Their eyes, their hair, their voices, honestly,
   That last word has an infidelic ring,
Were just the same, as was their daily dress—
In what they differed it was hard to guess.

For books and solitude did Olive crave,
   From early girlhood, and the taste still clung
To her maturer years; she was too grave
   For 'Manda's lighter moods, and oft among
The garden shrubs to reveries she gave
   Her afternoons, where lustrous orioles sung.
What wonder that she entered a convent,
When fair Amanda off to college went?

Five years of cloister life had its effect;
   She half forgot the little that she knew
Of worldliness, and so her self-respect
   Was still unshocked, when reaching Pleasant View,
Where lifted pastures, with the crocus decked,
   Invited her to roam their gulches thro';
She had no vanity, because an heir
With 'Manda, to a mansion old and fair.

AFFINITY.

By some odd, mysterious circumstance,
   Ingenuous Olive met a city swell,
Who boasted having studied art in France
   And had a few dry narratives to tell,
Whenever there occurred but half a chance;
   And Olive somehow grew to like him well.
He was a whole-souled fellow, with blue eyes,
That straightway won her deepest sympathies.

Our first love always carries us beyond
   The boundaries of rational esteem,
And Olive who had ne'er before been fond
   Of men, now made this one her constant dream.
With only rufiled sighs could she respond
   To Minet, when he wandered from love's theme,
But he returned and popped the question soon.
And they were married the next year in June.

The love, it proved, was all on Olive's side;
   Young Minet was a "flirt" and nothing more,
And yet he seemed to be quite satisfied—
   Perhaps, because he knew a goodly store

Of ducats was possessed by his fair bride.
  At any rate, he often deigned to pour
A few suspicions and soft-whispered dears
Into her innocent, attentive ears.

But Minet lived too fast and shortly all
  His money, like "Othello's job," was gone;
While plotting how to render light his fall,
  A brilliant thought, at length began to dawn,
Which caused his smile to grow majestical—
  His mind was fixed upon the terraced lawn.
The fountains, and the mansion old and fair,—
He told his wife that she should have her share.

The change was most agreeable all round,
  Especially to Olive, who, at night,
Had stayed awake, until the small hours found
  Her waiting still in cold and quaking fright
For Minet, who crept in without a sound,
  And ruthlessly blew out the welcome light.
Her convent life she much preferred to this—
She hated so the bad metropolis.

PART THIRD.

Amanda was surprised, but not less glad,
   That Olive with her worser half had come;
But she gave signs of feeling rather sad,
   And George appeared preoccupied and dumb.
The home machine was working very bad,
   'Twas out of gear, and needed mending some—
And all because, for fancy's bubbling sake,
These two had made a serious mistake.

  *   *   *   *   *   *

It was not strange that George, devoted yet,
   To science and to study, should discern
In Olive, one, whose character was set
   With precious gems, while ruefully in turn,
She saw, through eyes with keen repentance wet,
   An ideal man for whom her soul did burn.
Ah! bitter is the trial of woman's mind,
When faith is deep and innocence is blind.

  *   *   *   *   *   *

Then Minet and Amanda followed suit,
　　Discovering here and there a common trait;
She thought that he was beautiful and cute,
　　While he admired her dainty form and gait.
Transplanted love, they say, takes deeper root —
Without some bond 'tween mind as well as heart,
The husband and the wife will grow apart.

　*　　　*　　　*　　　*　　　*　　　*

Into the courts repaired this fine quartet,
　　And for divorces grievously applied;
But when the jury, twelve in number, met,
　　They stubbornly the applicants denied,
Which filled Amanda with a wild regret,
　　And Minet contemplated suicide;
While George half wished, despite all legal gyves,
That they could quietly exchange their wives.

　*　　　*　　　*　　　*　　　*　　　*

However, it is pleasant to relate,
    These sisters who appeared and dressed the same,
Submitted to the stern decree of fate,
    Thinking it best to suffer aught but shame;
They learned the worth of a congenial mate,
    By having none, that bore a husband's name.

*MORAL.

Whenever you have married life in view,
Be sure and choose some one that is like you.

---

* Similarity of taste rather than of temperament is meant.

SONGS AND VAGARIES.

THE CASCADE OF ROMANCE.

TO M. A. C.

Once in springtime it chanced that a maiden and I
  Went out on the beautiful Hudson to row;
There were blue-colored mists asleep in the sky,
  And a breeze from the hills fanned the ripples below.

When the city had faded I muffled my oar,
  For her charms took away all delights of the sail;
Absorbed in her words, we drifted to shore
  Where we heard a cascade up the moss-covered vale.

THE CASCADE OF ROMANCE.

On the rocks soon we stood, with the falls just above
   Where we watched woodland ferns by the bounding spray kissed;
Then and there I was dying to speak of my love—
   But an undefined feeling warned me to desist.

She appears to me now as she did in past hours
   When we drank from the cascade and gave it a name;
Her large wealth of blonde hair was brilliant with flowers,
   And the tint on her lips put the roses to shame.

Her beauty and tenderness still prompt devotion,
   Not to grow chill 'neath the footsteps of time;
Like a bark I may roam o'er love's treacherous ocean,
   With confidence still in this pilot of mine.

Ah! a dream rich with love seems to mellow my pen,
   And I oft fondly cling to a hope e'er denied;
To go to the Cascade of Romance again—
   With her all alone there in peace to abide.

REMORSE.

How much pleading would it cost,
    To restore those happy years,
To bring back the one I lost
    When my heart was full of pride;
    Pity then I had not died—
She so loving, trustful, kind,
I—a reckless youth and blind;
    Ever laughing at her tears.

But my conduct she weighed well,
    Though her eyes saw not my heart
Intuition broke the spell—
    Showed a nature low and mean
    Where no honor could be seen.
Self-reproached I madly swore,
I would never see her more—
    Since then we've lived far apart.

Other beauty have I known,
    Other virtue half divine;
But the old love dear has grown,
    With her sunny, flossy hair;
    It is now my only prayer,
Yet to see that happy day,
Brought about in love's own way
    When I e'er can call her mine.

LADY ROSE.

Lady Rose is rich, they say,
    But not vain;
She returned home yesterday—
    Home from Spain.
All the haunts she knew of yore,
Now will welcome her once more
    Where she'll reign.

Lady Rose is bland, they say,
    As the skies;
But her heart is light and gay
    Though it sighs.
What is beautiful and good
Finds a throne in womanhood
    Ere it dies.

Lady Rose is fair, they say,
    That I know—
For I met her one bright day
    Long ago.

Of her bloom remark I made,
Ah! this precious Rose will fade,
   Even so.

Lady Rose is loved, they say,
   Who could dare?
And her heart is far away;
   Tell me where!
I would roam the world around,
If that treasure could be found
   With her there.

*   *   *   *   *

Lady Rose's a bride, they say,
   Soon to be;
And she sails the first of May
   O'er the sea.
No saint is happier above,
Since Lady Rose gave her dear love
   All to me.

A BACHELOR'S STORY.

By a fountain which tinkled a silvery lay,
    She stood that night on the terraced lawn,
With her hand outstretched in the crystal spray,
    Which was fine as the gossamer dawn.

And her form was poised with an artless grace,
    As the angels have in Paradise;
And oh! the bloom on her Grecian face,
    And oh! the pride in her Roman eyes.

The shimmering light from a mellow moon,
    Was fading behind a sable cloud;
But the fountain still chimed a mystic tune
    To the dreaming ferns in their autumn shroud.

And I saw the beautiful girl still there,
    With her hand outstretched and her gaze above,
"A heart that is faint ne'er wins the fair,"
    Said I in a frenzy of love.

Then I cautiously crept to her lovely side,
    And pressed a kiss on her rosy(?) cheek;
Alas! it was cold; "art thou dead?" I cried;
    But a maid made of marble can't speak.

Well, I left her there and stole away
    From the statuette that deceived me so;
Her hand may still be in the crystal spray,
    And her form well poised for aught I know.

When the moon is obscured by a cloud above,
    I wander about the grounds alone,
And think of my first and my only love—
    That life-like maiden in stone.

THE SAME ANSWER.

If I knew the full import of omens that haunt me,
   I never could ask, love, if you are still true;
But sometimes the smile on your lips is so empty,
   That tempted I am to think strangely of you.
Long ago in the valley we wandered together,
   How sweet were the kisses I stole from you then;
But romance is subject to change like the weather,
   Oh, love, would you give the same answer again?

We paused on the bank of the foam-crested river
   To gaze on the tide as it drifted away;
And, darling, you know how your eyes seemed to quiver,
   As you waited to hear all my heart had to say.
Ere we crossed the old bridge on our way home together,
   The thing was decided, except as to when;
But romance is subject to change like the weather,
   Oh, love, would you give the same answer again?

THE SAME ANSWER.

The promise we had at the time of our marriage,
    Of wealth and position, is still unfulfilled;
In street cars we travel instead of a carriage,
    And trouble the roots of my gray hair has killed.
Now when you take all of my failures together,
    You often must think me the vilest of men;
Confess that your romance has changed like the weather,
    And that you'd ne'er give the same answer again.

Oh, dear wife, forgive me for doubting your passion,
    Your tear-flooded eyes your fidelity prove;
We'll live quite as happy if not in the fashion,
    And never again will I doubt your fond love.
Down the pathway of life we will journey together,
    Two lessons I've learned that may save me much pain;
That romance does not always change like the weather,
    And that you would give the same answer again.

HER NAME.

(FROM THE GERMAN OF LESSING.)

I once asked my sweetheart,
"What shall my grief call thee?
Wilt thou be Dorimene,
   Galathee or Chloris,
   Lesbia or Doris,
As the world knows her children?"

"Alas! names are only sounds to me,"
Quoth my true love bitterly,
   "Thou may'st choose it, call me Doris,
   Or Galathee, or Chloris,—

*   *   *   *   *   *

"A sweeter name than those or mine
Would be to call me only thine."

LOVE'S SILENCE.

Some one is waiting for the moon to rise
And shed its lustre o'er autumn skies;
    Some one is waiting in twilight's dell
    For one she expects who loves her well.
Sighing, perchance, at the mottled void
Of endless hopes and fears alloyed;
    She ne'er can judge of the changing view
    Until she knows that all is true.

So, nervously plying the thread of thought
Thro' all his words and actions brought,
    What can the lady do but meet
    Their author bowing at her feet,
In whom so many feelings brood,
That none can well be understood?

*   *   *   *   *

In this hushed hour, with passion pent,
Love smiled when silence gave consent.

A MAIDEN LADY.

Should curses tinge to deeper rosiness
   Her lips, now ripe with aged innocence,
I could not summon one rebuke, but bless
      Their harsh laments.

For she has fostered all the bliss of love,
   That any virgin heart could well possess,
And she has prayed to God that he remove
      Its bitterness.

'Tis hard for womankind to bear the pain
   Of wounded love, of man's deceitful smile:
She's never fit to love or bear again
      Another trial.

Some women rave and mock love's aftermath
   When once their chastened faith in men is hurt:
But Theresa only sighed, she showed no wrath —
      Nor could she flirt.

A MAIDEN LADY.

Untutored in the art of intrigue, she
    Still lived on, hoping for love's brighter dawn—
Not knowing aught of life's dark mystery
    When love is gone.

And she was loyal still to her heart's choice,
    To him who wrung from it a holy vow;
Still true to him, whose lost and far-off voice,
    She answers now.

REDEEMED.

Sob on, dear heart, if grief
Finds in sobbing a relief,
In freeing long-pent waters of the heart;
Yet had these tears been kept
In your sweet eyes still unwept,
Your enemies had know one triumph less.

I know 'tis true, Corinne,
The world has only seen
That one false step you made long years ago;
And never till you're dead,
Will it here on earth be said,
How kind you were to God's poor ones, Corinne.

How you have made hearts glad,
When your own was bleak and sad,
How you won from them a tearful gratitude;
Ah well! poor heart, some day
When Corinne is laid away,
All will know you paid an honest debt to God.

REDEEMED.

You were beautiful and young,
Envy aimed her darts that stung
Like a serpent, ere the world had seen your fall;
And I came to you too late
To save you from their hate;
But not too late to say, "repent, Corinne!"

Ah, yes; your friends are few;
But, Corinne, there's one still true,
And I know of more, but they are up in heaven;
You will not regret the dearth,
Of the friends who live on earth,
When you rest within the Sacred Heart forgiven.

Hush now, sob not to-day,
For the stain is washed away
In your own repentant passion-drowning tears;
Since your peace with Him is made,
All the gloomy past will fade,
And you'll bathe God's feet with tears of joy,
Corinne.

WRITE OFTEN TO THE OLD FOLKS.

"Write often to the old folks,"
   Said sister May to me,
"You're going off to college, Will,
   'Mong strangers you will be;
I know you'll work as well as play,
   But whatsoe'er you do,—
Please don't forget that we, at home,
   Will long to hear from you.

"Now, Will, don't think me foolish,
   But mother is not strong,
And she will surely worry
   If you put off writing long.
I know her eyes would sparkle,
   And a bloom would tinge her cheek—
If you could only write her
   A letter once a week.

"And father can advise you
   If anything goes wrong;
Write him about your troubles,
   For he is wise and strong,

WRITE OFTEN TO THE OLD FOLKS.

Guard well your habits, brother,
   And when back from school you come,
You will find a hearty welcome
   From the cherished ones at home.

"Write often to the old folks,
   Their hair is growing gray,
Not very many years, alas!
   Have they on earth to stay.
Oh, promise me this favor,
   And never will you rue
The day you write the old folks,
   Who will long to hear from you."

"Dear sister, this I promise,"
   And my tears began to flow—
"I'll write often to the old folks,
   If you think 'twill please them so."

*   *   *   *   *

And I write a weekly letter,
   In my snug and cheerful room—
And sister May informs me:
   "Mother's cheeks are in full bloom."

AN IDEAL HOME.

OF A POET WHO IS MARRIED BUT NOT UNHAPPY.

Oh. give me a home where the Muses may come,
   And bring me their tokens and love;
Let it be far away from the prosaic hum,
In a clime where the hammer of labor is dumb—
   In the midst of a sweet olive grove.

And there with Euterpe and Thalia alone,
   The ages will seem but an hour,
As I sit on my highly poetical throne,
And hear every song and each nectarous tone
   Which will be like the song of a flower.

But the rest of the Nine would be welcome there too,
   With the Graces and fairies and elves;
The former could keep all their dignity true,
The latter could romp, give me kisses—a few,
   And whisper their dreams 'mong themselves.

AN IDEAL HOME.

More welcome than all to this Eden retreat
  Is the one chief concern of my life;
The one without whom it would be incomplete,
And 'reft of a presence immeasurably sweet—
  It's my lovely but tangible wife.

A PEACE OFFERING.

I would weave you a song just as sweet as the flowers
    Which exhale a heart message from you;
But I fear I am lacking the requisite powers
    To do what Lord Byron could do.

So, if you'll excuse a soft, languishing lay,
    And accept a plain common sense rhyme,
I'll indite a few lines and endeavor to say
    That I'll be your true friend from this time.

But it is not because I'm in need of a friend
    That I'm asking you now to be mine;
For to tell you the truth I have plenty to lend—
    At their absence I scarcely could pine.

In grateful remembrance of what you have done
    To a pitiful object like me;
I frankly confess that my heart you have won,
    And at your disposal 'twill be.

A PEACE OFFERING.

If ever we meet on this cold hemisphere,
   Don't let me forget to improve
The occasion by calling you softly "my dear,'
   And you answer fondly "my love."

MARASCHINO.

Without, in the distance, the snow gleams white,
  'Neath the pale, cold moonlight the frost leaves
    twine;—
Within, there is color and warmth and light,
And song—all so pleasant to sound and sight—
The fragrance and glow of a summer night
And my love in the midst of it fair and bright—
  Dipping her lips in the wine.

Just over the table my darling sits,
  Above her the warm lights shimmer and shine;
She smiles, then a frown her fair brow knits,
While over her pale cheek a rose flush flits;
She's enough to bewilder the gravest wits,
And I—ah! I love her—my queen who sits
  Dipping her lips in the wine.

At first she looks bland and softly sighs,
  With glimpses of laughter her dark eyes shine;
I can see each merry conceit arise,

And glimmer and ripple until it flies
From the dimpled cheek and the roguish eyes,
Where mischievous Love in ambush lies—
   To the lips that are sipping the wine.

And dreamily over her creeps the spell,
   And dreamy and tender the dear eyes shine ;
No more of laughter their glances tell,
And silence falls as of old it fell
In Paradise—where Love came to dwell—
Not silence that severs but binds us well,
   As she dips her lips in the wine.

The light that I look for by her unguessed,
   Steals into her eyes as they turn to mine ;
One little white hand stays out in quest—
With fluttering pulse too glad to rest—
Of another hand, where it's caught and prest.
Ah ! more red than the blossom she wears on her
     breast
   Are the lips that are sipping the wine.

Just one little instant with joy replete,
   As her lovely eyes upturn to mine;
The moment of moments, so perfect, complete,
While I briefly gaze on her figure neat,
And the flowers' faint perfume my senses greet;
I touch—or I try to—but they are too fleet—
   The lips that were dipped in the wine.

LEVITIES.

SONNET.

TO AN ORGAN GRINDER.

Here, stranger, take my penny for the tune
    That from thy stone-scared box hath deftly rolled;
    Like yonder hills the strains are rather old—
They don't remind me of the songs of June;
But any kind of music is a boon
    In this dull round of silence, so please grind
    Another sprightly waltz, to ease my mind
E'er brooding on some ebon theme and cold.
Be quick, before arrives that crowd of boys!
    Thou art a swarthy being, but thy rank
As a musician brings thee varied joys,
    And many coppers to thy savings-bank.
Go elsewhere, now, with all thy lonesome noise—
    Or thou wilt make me like thy haft—a crank.

TRIOLETS.

You should have gone before,
 The rain's falling faster,
But you said one kiss more;
You should have gone before,
 You are an awful laster.
You should have gone before—
 For the rain's falling faster.

Oh! Jane, I could not go,
 Your lips taste so like honey;
You want me here you know,
And, Jane, I could not go;
 I would not leave for money.
Oh! Jane, I could not go,
 Your lips taste so like honey.

Now, Tom, you cannot stay;
 I thought I heard my father—
To gnaw my lips that way,

TRIOLETS.

Now, Tom, you cannot stay ;
   You are a dreadful bother—
Now, Tom, you cannot stay,
   I'm sure I hear my father.

Farewell, forever, Jane,
   Out thro' the mud I'll scour ;
My heart is full of pain,
Farewell, forever, Jane,
   Your kisses are too sour ;
Farewell, forever, Jane—
   Out thro' the mud I'll scour.

ONLY COUSIN SAM.

Oh! he was such a charming beau:
With voice so musical and low;
And you would think him just 2, 2,
Although he was not twenty, too—
    He was my cousin Sam.

Once in the gloaming he and I
Sat talking; fast the hours slipped by,
And soon 'twas midnight I declare;
But still we lingered fondly there.
    'Twas only cousin Sam.

He told me all his rap'rous love,
He called me his seraphic dove—
Nay, placed his arms about my waist,
He kissed me—said he liked my taste.
    'Twas only cousin Sam

He asked me for my heart and hand
In words I could not understand;
But just to make my lover calm,
I said I didn't give a—cent,
    Since it was only Sam.

THE REASON.

HE.

What storm is brewing in my lady's breast,
That angry frowns should dart across her brow
    And tear-drops flood the heaven of her eyes?
What wild unshackled tempest of unrest
Chafes into foam the nectar on her lips
    And lashes into fury all her sighs?

SHE.

'Tis no mere storm that's raging in my heart,
It is a cyclone, sweeping thro' my soul,
    Destroying love's bright temple as it goes.
It rends our matrimonial ties apart,
And all because you are so miserly
    That you won't buy me any satin hose.

IN 1901.

There'll be a new drink for the people to quaff
    In 1901.
There will be some new jokes at which we can laugh
    In 1901.
Jay Gould will not manage the telegraph
    In 1901.
Tom Thumb and Jumbo will sleep in the tomb
    In 1901.
The sunflower and lily will not be in bloom
    In 1901.
Mr. Wilde will be found in a dark garret room
    In 1901.
The electric light will turn night into day
    In 1901.
'Twill not be thought naughty to go to the play
    In 1901.
And many a husband will have his own way
    In 1901.

IN 1901.

There'll be no rum license down there in Maine
    In 1901.
In Congress again may appear Mr. Blaine
    In 1901.
Those who are dead will not feel any pain
    In 1901.
There's a rumor afloat that "the Chinese must go"
    In 1901.
And that Mr. Hanlon refuses to row
    In 1901.
But the "drummer" and whale will continue to blow
    In 1901.
There'll be some new converts to Darwin's cause
    In 1901.
Each State will be willing to change all its laws
    In 1901.
To watch our big country the whole world will pause
    In 1901.
The girls will like taffy. Oh, just the same,
    In 1901.
And General Grant will be known to fame
    In 1901.

He'll die a third termer in Liberty's name
    In 1901.
Mary Walker, Esq., will not pant for praise
    In 1901.
There'll be nothing at all to the crockery craze
    In 1901.
Old people will yearn for the "good old days"
    In 1901.
Oh! who will dare wear the Grecian bend
    In 1901?
And which of us now will have money to lend
    In 1901?

  *   *   *   *   *   *

I fear this bad world will come to an end
    In 1901.

www.ingramcontent.com/pod-product-compliance
Lightning Source LLC
Chambersburg PA
CBHW021947160426
43195CB00011B/1264